ZERO!

The Number That Almost Wasn't

Sarah Albee

Illustrated by **Chris Hsu**

ini **Charlesbridge**

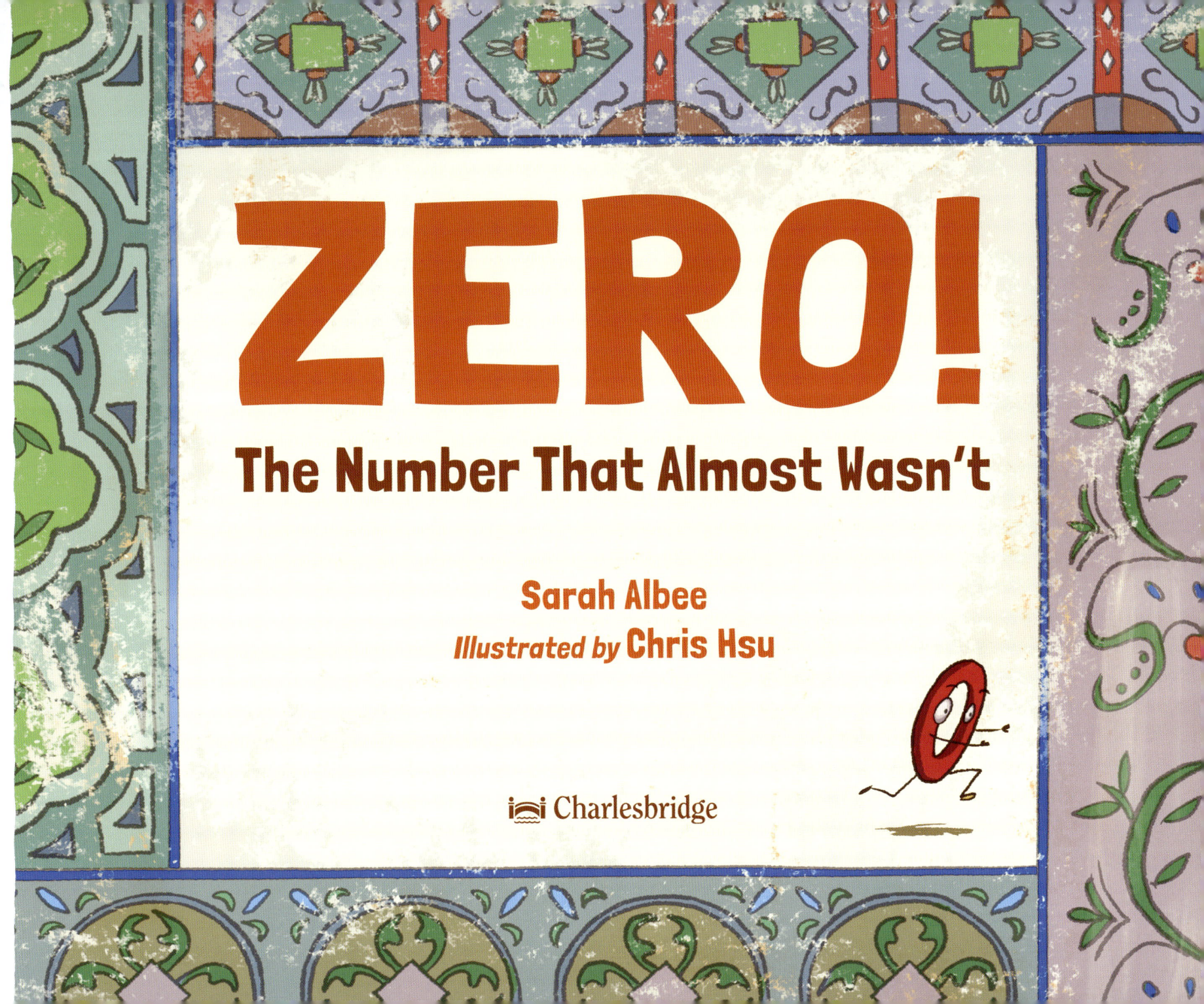

We all know what *nothing* means.
But how often do you *really* think about zero?
After all, it's a number that means nothing.

It's always been there—

Zip!

Nil!

on clocks,

Zilch!

thermometers,

and math homework.

Diddly- squat!

C° — F°
100 — 212
90 — 194
80 — 176
70 — 158
60 — 140
50 — 122
40 — 104
30 — 86
20 — 68
10 — 50
0 — 32

100.00

0

$$\begin{array}{ccc} 10 & 10 & 18 \\ \times 5 & \times 7 & +12 \\ \hline 50 & 70 & 30 \end{array}$$

$$\begin{array}{ccc} 35 & 20 & 56 \\ -15 & \times 2 & -26 \\ \hline 20 & 40 & 30 \end{array}$$

(100) A+

0

0

Right?
Well . . . not quite.
Zero had to be invented.

Early humans counted on their fingers.
To keep track of calculations,

they made notches on sticks,

or piled up pebbles,

or tied knots in strings.

Over time, people invented written symbols for numbers so they could keep track of how many sheep they sold or bundles of grain they traded. But no one thought to invent zero.

Of course early humans understood the concept of nothing: if you have six figs and eat them all, you have no figs left.

But if your basket is empty, how can you count what isn't in it?

Why create a symbol to represent nothing?

Zero wasn't an obvious idea.

Then cities grew larger. Merchants sold more stuff. People paid taxes.

Someone needed to come up with a better way to calculate.

About four thousand years ago, the Babylonians did just that.

They invented place value!

The position of each digit in a numeral tells us its value.

7   3   2   6

In this position the digit 7 has a value of 7,000.

In this position the digit 3 has a value of 300.

In this position the digit 2 has a value of 20.

In this position the digit 6 has a value of 6.

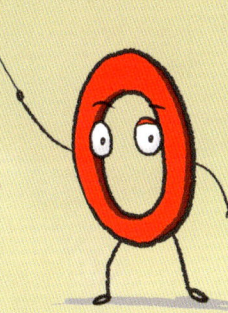

So the numeral 7,326 is made up of 7,000, plus 300, plus 20, plus 6.

But the Babylonians didn't have a symbol for zero. With no zero, how could they write a numeral like 305, with nothing in the tens place? Or 2,035, with nothing in the hundreds place?

And how could anyone tell the difference between 16 and 106 and 1,006?

Finally a clever Babylonian scribe came up
with a solution: Why not put a special mark
in the empty place?

The Babylonians created a symbol for an
empty place. It was a start!

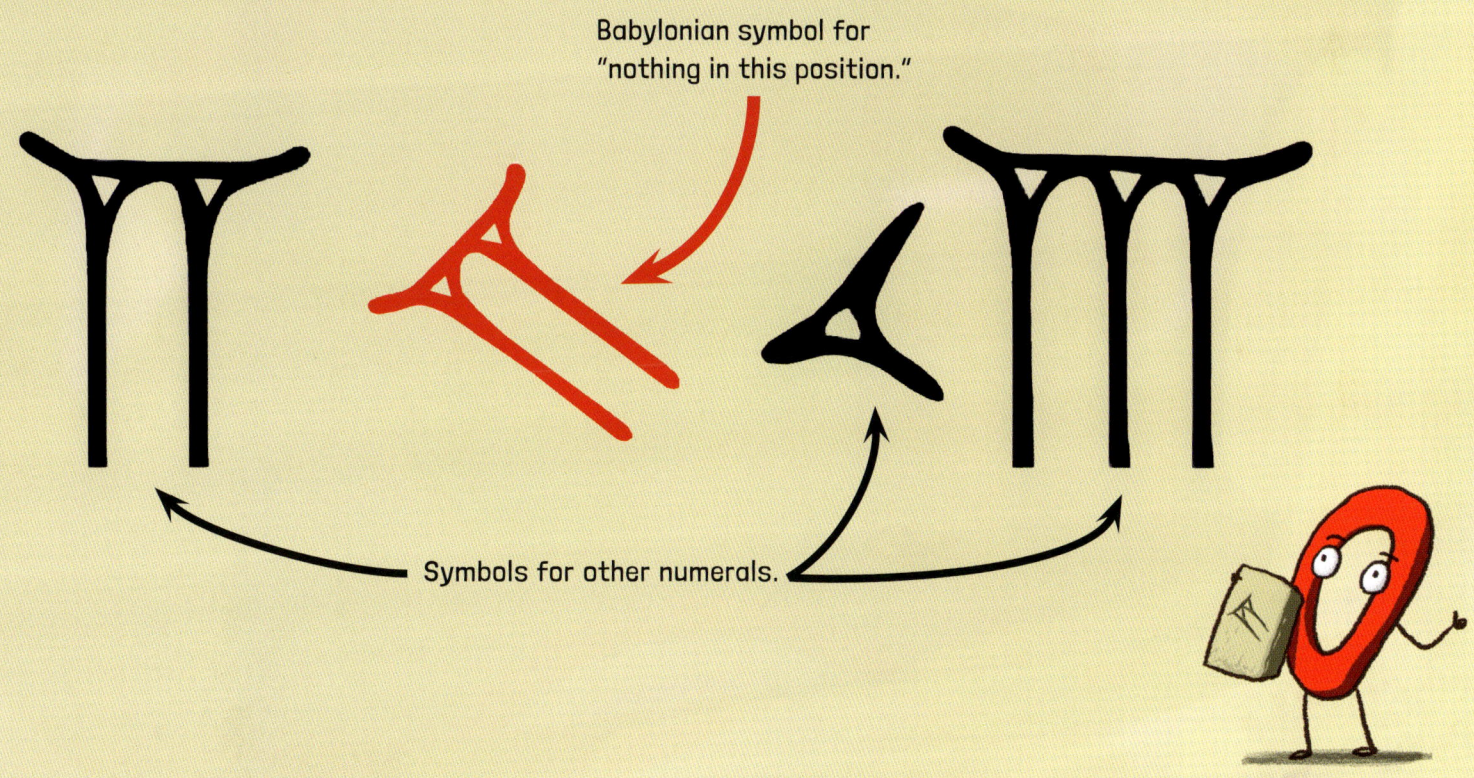

Babylonian symbol for
"nothing in this position."

Symbols for other numerals.

Meanwhile in ancient Greece, astronomers and mathematicians excelled in geometry—the math of shapes and spaces. But the Greeks had a clumsy system that used letters to represent numbers.

For basic calculations, Greek merchants either counted on their fingers or used a special tool known as a counting board.

And they didn't have a symbol to represent nothing.

A few centuries later, the Maya invented a symbol for zero. They chiefly used it in their complex calendars.

The Maya zero looked like this.

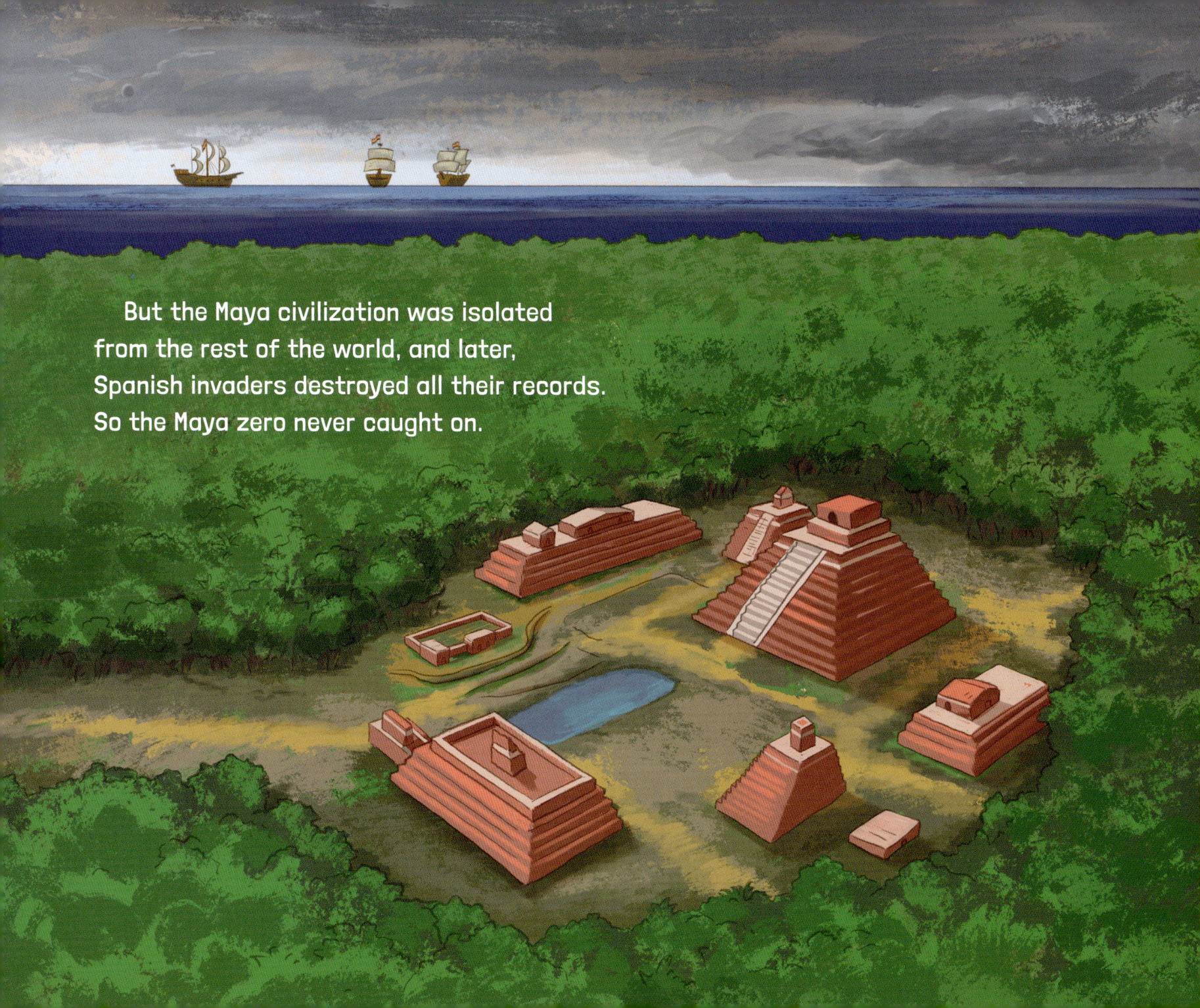

But the Maya civilization was isolated
from the rest of the world, and later,
Spanish invaders destroyed all their records.
So the Maya zero never caught on.

Around the same time that the Maya civilization was flourishing, the powerful Roman Empire was conquering much of Europe.

The ancient Romans had no zero. And they had no place-value system.

To perform basic calculations, they used their fingers, pebbles, or a counting board.

Then they recorded their numbers using seven letters: M, D, C, L, X, V, and I.

After the Roman Empire collapsed, much
of Europe entered a dark time. Most people
couldn't read or write. Wars were frequent.
And still no one thought to use zero.

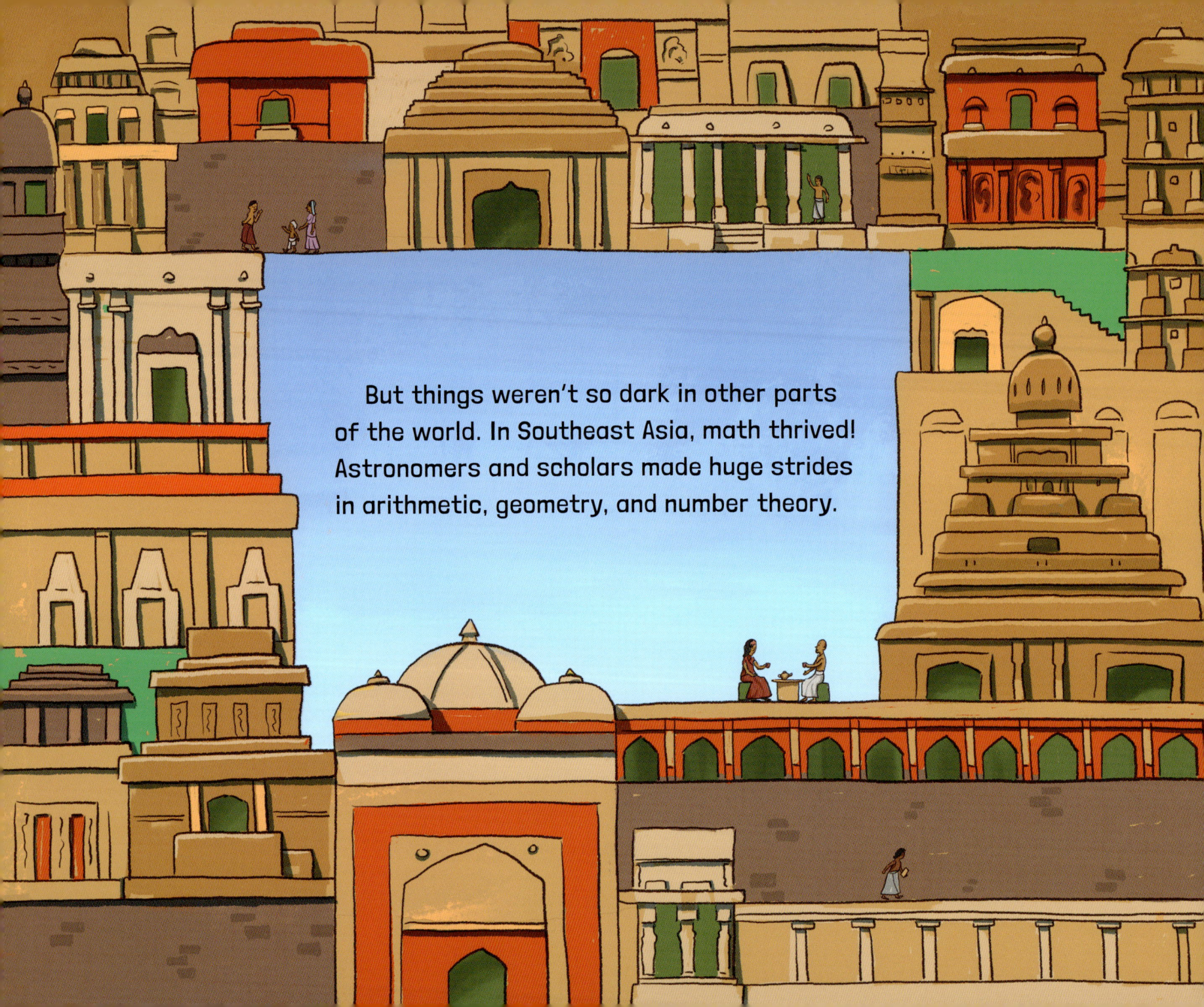

But things weren't so dark in other parts of the world. In Southeast Asia, math thrived! Astronomers and scholars made huge strides in arithmetic, geometry, and number theory.

And then one day an unknown Indian mathematician made a small dot on a piece of birch bark.

They created a symbol for zero and used it to write down numbers!

A few centuries later a brilliant Indian mathematician named Brahmagupta wrote a book, in Sanskrit, that explained the ten Indian number symbols and place value and how to calculate with zero.

Constantinople

Egypt

Baghdad

Traders introduced zero and the new numeral system to China and parts of Southeast Asia. Zero also traveled west to Baghdad, the bustling center of the Muslim Empire.

A great Persian mathematician named Muhammad ibn Mūsā al–Khwārizmī wrote a book in Arabic about the Hindu (Indian) numerals, which introduced zero to that part of the world.

Al–Khwārizmī wrote another book about a method traders used to solve arithmetic problems. Today we call it algebra.

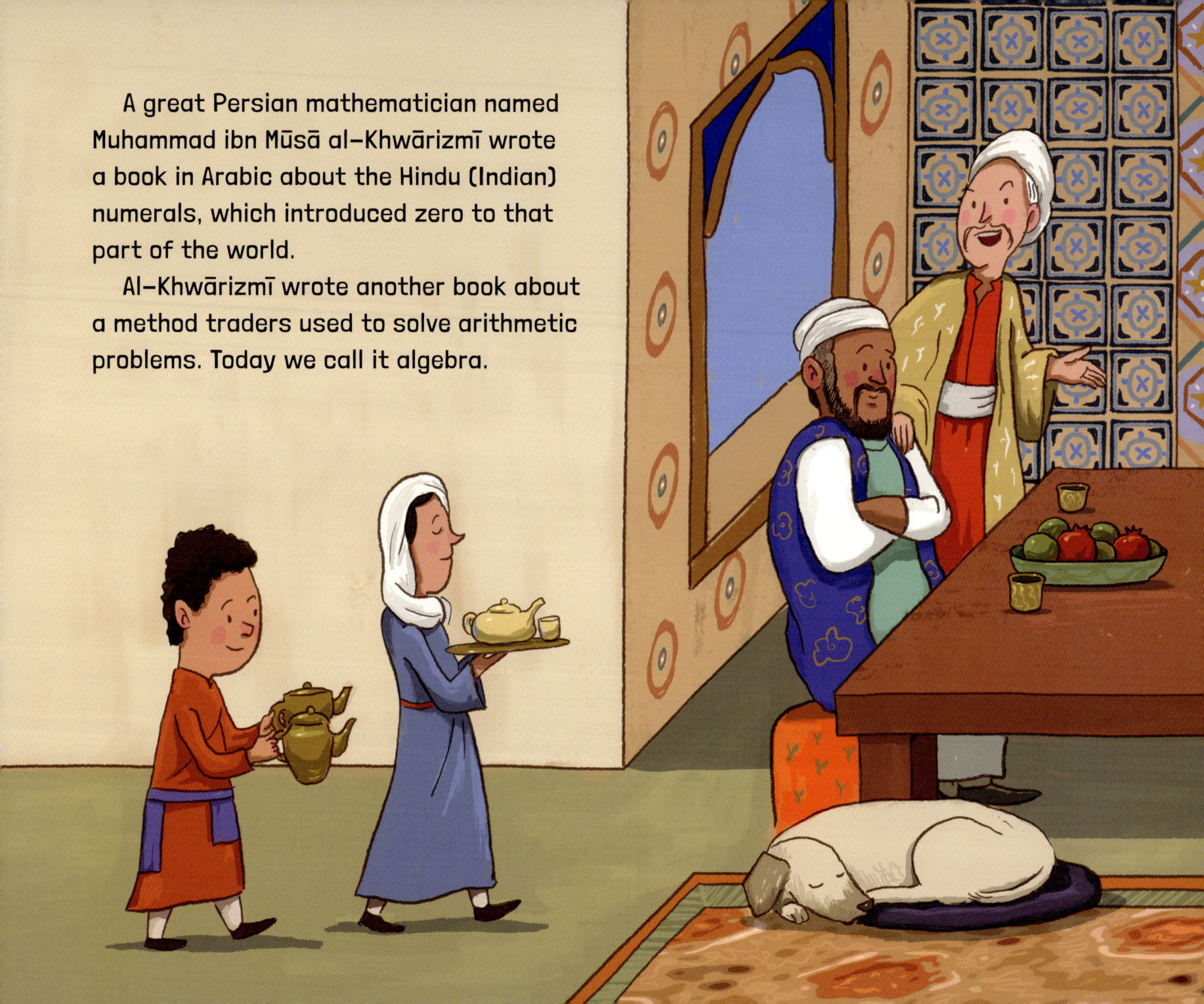

Traders brought books written in Arabic by al-Khwārizmī and other mathematicians to Europe. These ten numerals—0, 1, 2, 3, 4, 5, 6, 7, 8, and 9—would become known as Hindu-Arabic (or Indo-Arabic) numerals.

A few European mathematicians enthusiastically embraced the idea of zero.

But many others were suspicious of the newfangled invention. They decided that zero was a dangerous, foreign idea created by people they didn't understand.

An Italian merchant learned about the Hindu–Arabic number system while traveling in North Africa. His name was Leonardo of Pisa, although we call him by his nickname: Fibonacci.

Fibonacci happened to be an excellent mathematician. He wrote his own book, in Latin, that explained how to calculate with the new number system.

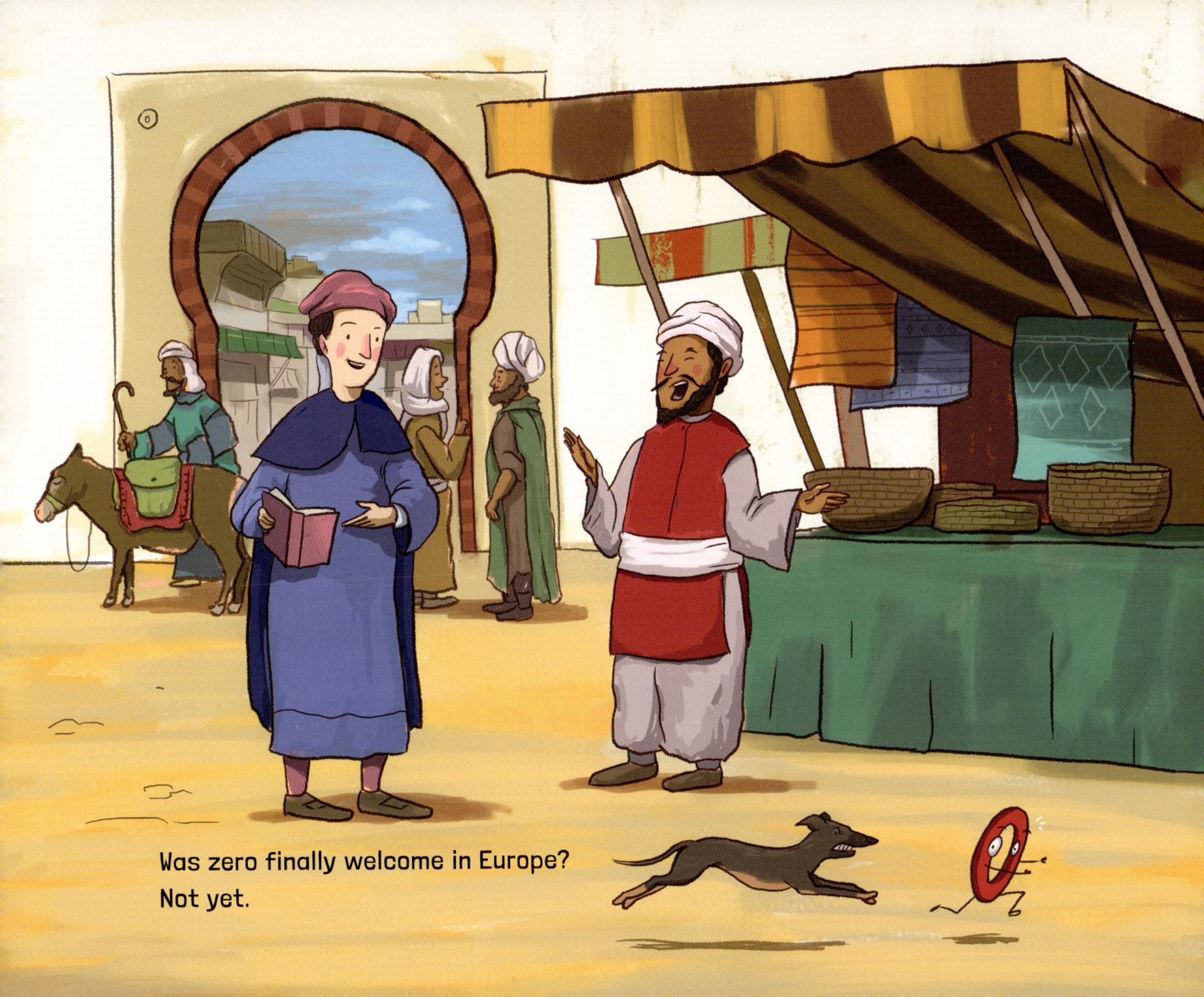

Was zero finally welcome in Europe?
Not yet.

Fibonacci's book made him famous throughout Italy, and some merchants started to use the elegant Hindu–Arabic number system in their daily trading.

But many people continued calculating with counting boards and writing down the results with Roman numerals.

A few Christian leaders actually *banished* zero. They argued that God had created everything, so something that represented *nothing* must be the work of the devil.

But in time more and more people made the switch.

What finally changed people's minds?
The invention of the printing press helped!

More printed texts became available, and more and more people learned to read. As a result, more and more people learned about the Hindu–Arabic number system and grew to appreciate its beautiful simplicity.

With the ten number symbols, ordinary people could finally perform basic calculations. And mathematicians could ponder ever more complex ideas.

Isaac Newton

Two great thinkers invented a new field of math called calculus. And that helped spark a scientific revolution.

Gottfried Wilhelm Leibniz

SCIENCE THEATER

EARLY HUMANS

CONSTELLATIONS

NOW SHOWING

NEWTON vs LEIBNIZ

TODAY'S SHOWTIMES!
10:00
11:00
12:00

HALLEY'S COMET

CLICK FOR MORE ▶

When people all over the world at last accepted zero, new fields of knowledge opened up:
modern physics,
electronics,
engineering,
computers.

Today zero is an essential part of our number system.

It took thousands of years and a few brilliant thinkers to imagine a world with zero. And now it's hard to imagine a world *without* it.

What's in a Name?

Zero

In Sanskrit (an Indian language), the word for zero is *śūnya*, which means nothingness. In Arabic, that word is *sifr*. We get the word *cipher* from that Arabic word. Nowadays *cipher* is an old-fashioned English word, but to *cipher* can mean to do arithmetic, and a *cipher* can mean a secret code. Later the Arabic word *sifr* was translated to the Latin word *zephirum*. Eventually that word evolved into zero.

Today we use lots of words that mean nothing. *Nil* and *null* have been around since the fifteenth century and have roots in English, French, and Latin. *Zip, zilch*, and *diddly-squat* have unknown origins and were first used in the twentieth century.

In sports, when a team scores zero points, people sometimes say, "The team scored a goose egg," because an egg's shape looks like zero. The French word for egg is *l'oeuf.* That may be why a zero score in tennis is called *love*. *L'oeuf* and *love* sound alike.

British people sometimes say "nought" or "naught" to mean zero. In the United Kingdom, the game tic-tac-toe is called noughts-and-crosses. The word *naughty* originally meant poor (having naught, or nothing), but evolved over time to mean untrustworthy or disobedient. That change might reflect many Europeans' long-standing mistrust of the concept of zero.

Some Definitions

Number: A value—count, measurement, or quantity—that can be expressed in many ways. For example, the number four can be expressed as the symbol 4, the word *four*, clapping four times, and more.

Numeral: A symbol or a name that stands for a number. For example, 12 and twelve are both numerals. So the number is the value, and the numeral is how we write it. Nothing is a value, and 0 and zero are the numerals we use to express the idea of nothing.

Digit: A symbol used to make a numeral. The word *digit* means finger in Latin. In the Hindu-Arabic number system, there are ten digits: 0, 1, 2, 3, 4, 5, 6, 7, 8, and 9. With these ten digits we can represent any number. So digits make up numerals, which stand for values (numbers), just the way letters make up words, which stand for ideas.

Hindu-Arabic Number System

Today we can use the ten-digit Hindu-Arabic number system to represent any number, which is a huge improvement from earlier systems. Although the symbols we use for the ten numerals have changed over time, they make up what's called a base-10 system. The Babylonians used a base-60 system, and the Maya used a base-20 system. Many computers use a base-2, or binary, system that uses only two numerals: 0 and 1.

Nowadays almost every country in the world uses the Hindu-Arabic base-10 system, which wouldn't be possible without zero.

Map Names

The names of many cities and countries have changed since ancient times. Here and throughout this book, we've used modern names unless otherwise noted.

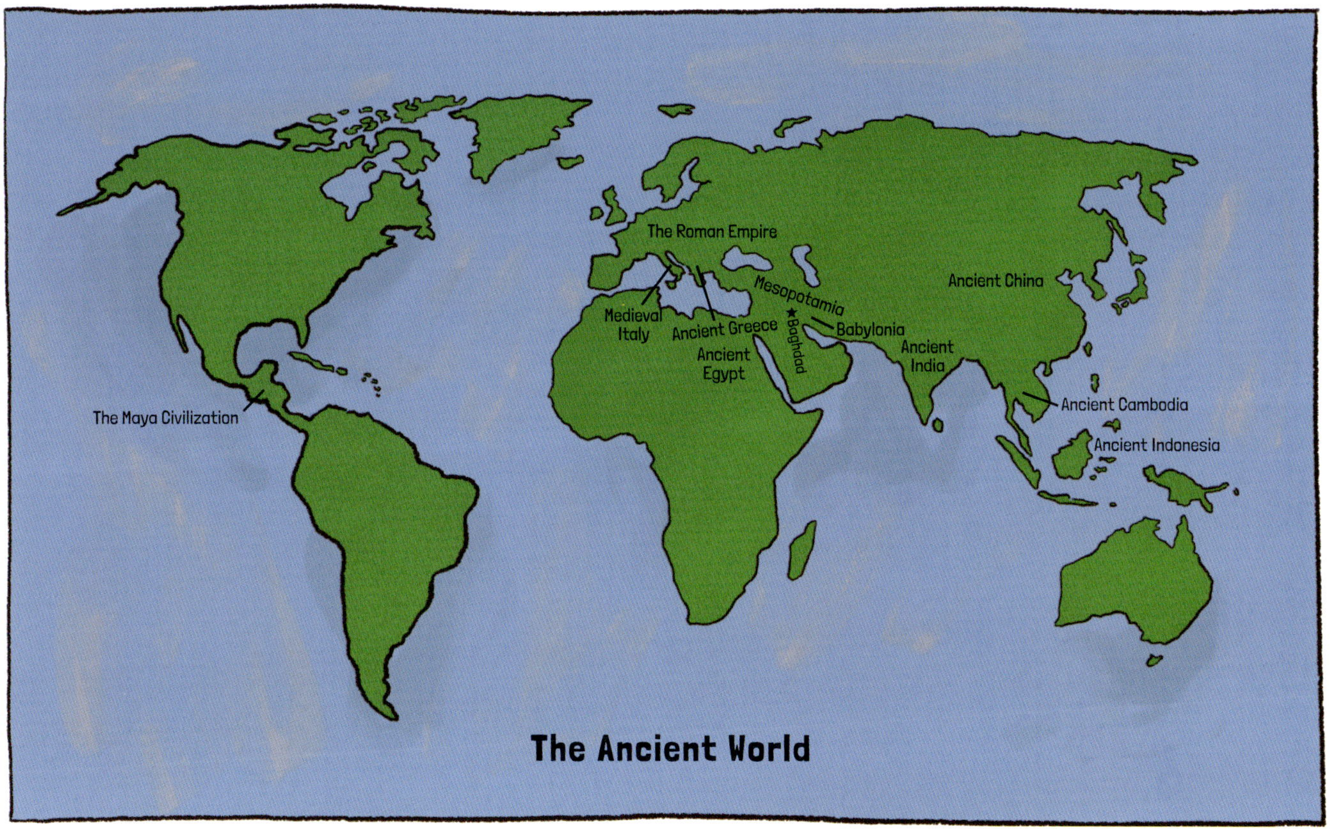

The Roman Empire

Mesopotamia

Ancient China

Medieval Italy

Ancient Greece

★ Baghdad

Babylonia

Ancient Egypt

Ancient India

Ancient Cambodia

Ancient Indonesia

The Maya Civilization

The Ancient World

Selected Bibliography

Aczel, Amir D. *A Strange Wilderness: The Lives of the Great Mathematicians*. New York: Sterling, 2011.

Al-Khalili, Jim. *The House of Wisdom: How Arabic Science Saved Ancient Knowledge and Gave Us the Renaissance*. London: Penguin Books, 2012.

Devlin, Keith. *The Man of Numbers: Fibonacci's Arithmetic Revolution*. New York: Walker Books, 2011.

Ifrah, Georges. *The Universal History of Numbers: From Prehistory to the Invention of the Computer*. Translated by David Bellos, E. F. Harding, Sophie Wood, and Ian Monk. New York: John Wiley, 2000.

Kaplan, Robert. *The Nothing That Is: A Natural History of Zero*. Oxford: Oxford University Press, 1999.

Seife, Charles. *Zero: The Biography of a Dangerous Idea*. New York: Penguin Books, 2000.

Books About Math for Younger Readers

Adler, David A. *Place Value*. New York: Holiday House, 2016.

Barton, Bethany. *I'm Trying to Love Math*. New York: Viking Books for Young Readers, 2019.

LaRocca, Rajani. *Bracelets for Bina's Brothers*. Watertown, MA: Charlesbridge, 2021.

LaRocca, Rajani. *Seven Golden Rings: A Tale of Music and Math*. New York: Lee & Low Books, 2020.

Stuart, Colin. *The Language of the Universe: A Visual Exploration of Mathematics*. Somerville, MA: Big Picture Press, 2020.

Zero: A Blurry Time Line

There's a lot about zero's history that's uncertain. Historians continue to debate and readjust dates as they uncover new information.

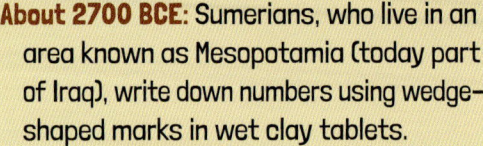

About 2700 BCE: Sumerians, who live in an area known as Mesopotamia (today part of Iraq), write down numbers using wedge-shaped marks in wet clay tablets.

About 400 BCE: The Greeks are aware of the concept of zero, but the philosopher Aristotle notes that it is impossible to divide by zero. So the Greeks don't consider it a number and don't use it. (Try dividing any number by 0 on a calculator. It will show that it can't be done.)

About 300 BCE: Babylonian tablets show zero as a placeholder. The concept spreads to India.

About 34 BCE: The Maya, who live in what is later called Central America, develop zero. But because there are oceans between their civilization and those in Asia and Europe, the Maya's zero has no influence on other numbering systems.

Somewhere Around 200–400 CE: The first written zero used as a number (rather than merely a placeholder) appears in a manuscript written on birch bark in ancient India. It's now known as the Bakhshali manuscript, and radiocarbon dating places it in the third or fourth century.

476: The Roman Empire collapses. Zero remains unknown in Europe.

628: Brahmagupta (598–670) writes a book (called Brāhmasphuṭasiddhānta) that defines zero as a number. His text may be the first to use zero in mathematical calculations.

683–685: Several stone tablets show the use of zero in what are now the countries of Cambodia and Indonesia.

Sometime in the 700s: Zero appears in China.

762: Baghdad (in what is now Iraq) is founded as the center of the Muslim Empire.

About 773: The concept of zero reaches Baghdad.

825–830: Mathematician Muhammad ibn Mūsā al-Khwārizmī learns about Brahmagupta's book and writes his own books. One is about calculating with Hindu numerals. Another introduces a method for solving arithmetic problems. Its title in Arabic is *Kitāb al-jabrawa'l-muqabāla*. Today we use a version of the "al-jabr" part of his book title and call that method algebra.

Sometime in the 800s: Zero spreads to other Arab territories, including Moorish Spain and North Africa.

1095–1291: The Muslims and Christians fight holy wars known as the Crusades. They are suspicious of each other's cultures, which might explain why Christian Europeans are slow to accept zero.

Sometime Between 1150–1175: An Italian scholar named Gerard of Cremona translates al-Khwārizmī's writings into Latin. Al-Khwārizmī's name is translated as "Algoritmi," which is where the English word algorithm comes from.

1200s: The merchant and mathematician Leonardo of Pisa (Fibonacci) publishes a book that helps popularize the 0 to 9 base-10 number system used today.

1300s: Europeans begin replacing Roman numerals with Hindu-Arabic numerals. In some European countries, using the new numerals in official documents is forbidden, in part because people think it's easier to forge them than Roman numerals. (They point out that a 0 can easily be changed to a 6 or a 9.) Or possibly they don't want to admit that the Arabs and Indians invented a much better system.

1300s–1400s: A printing press is invented in Korea in the 1300s, and another in Europe in the 1400s. People all over the world have better access to printed information. More people learn to read and to calculate with numbers.

1660s: Two mathematicians—an Englishman named Isaac Newton and a German named Gottfried Wilhelm Leibniz—independently develop a new kind of math called calculus, which sparks all kinds of modern inventions.

1689: Leibniz creates the first binary system—a series of zeroes and ones—which converts verbal language into math. This becomes a basis for modern computers and electronics.

1700s: Zero, part of the Hindu–Arabic base–10 number system, finally becomes widely accepted around the world, although many European banks and institutions use Roman numerals for a while longer.

Notes About the Art

- **Early humans:** The sheep are awassi, one of the oldest breeds in the Middle East.
- **Ancient Greeks:** Halley's Comet is in the sky because ancient people saw it, too.
- **Maya:** A solar eclipse is partially visible in the clouds above the left ship. There was an eclipse while making this book, and it was a bad omen back then.
- **Birch bark manuscript:** The mango is India's national fruit.
- **Persia:** The dog is a saluki, an ancient breed.
- **Fibonacci:** The dog is a sloughi, a breed developed in North Africa in the thirteenth century. The artist's composition is roughly organized in the Fibonacci sequence.

- **Printing press / Italian Renaissance:** The Duomo (an Italian cathedral) is under construction (back). The Florence Baptistery is complete (left). The crowd includes a couple from *The Diptych of Federico da Montefeltro and Battista Sforza* by Italian artist Piero della Francesca.
- **Isaac Newton:** Yes, he was left-handed!

Text copyright © 2025 by Sarah Albee
Illustrations copyright © 2025 by Chris Hsu

At the time of publication, all URLs printed in this book were accurate and active. Charlesbridge, the author, and the illustrator are not responsible for the content or accessibility of any website.

Charlesbridge · 9 Galen Street, Watertown, MA 02472 · www.charlesbridge.com

Library of Congress Cataloging-in-Publication Data
Names: Albee, Sarah, author. | Hsu, Chris, illustrator.
Title: Zero! the number that almost wasn't / Sarah Albee; illustrated by
   Chris Hsu.
Other titles: Zero!
Description: Watertown, MA: Charlesbridge, [2025] | Includes bibliographical
   references. | Audience: Ages 7–10 | Audience: Grades 2–3 | Summary:
   "The history of the number zero is long, complicated, and interesting."
   —Provided by publisher.
Identifiers: LCCN 2023056533 (print) | LCCN 2023056534 (ebook) |
   ISBN 9781623544324 (hardcover) | ISBN 9781632893918 (ebook)
Subjects: LCSH: Zero (The number)—Juvenile literature. | Cardinal numbers—
   Juvenile literature. | Mathematics—History—Juvenile literature.
Classification: LCC QA141.3 .A43 2025 (print) | LCC QA141.3 (ebook) |
   DDC 513.5—dc23/eng/20240326
LC record available at https://lccn.loc.gov/2023056533
LC ebook record available at https://lccn.loc.gov/2023056534

Printed in China · OPIC
The authorized representative in the EU for product safety and compliance
   is eucomply OÜ Pärnu mnt 139b-14, 11317 Tallinn, Estonia,
   hello@eucompliancepartner.com, +33757690241
(hc) 10 9 8 7 6 5 4 3 2

Illustrations created in Adobe Photoshop
Text type set in Londrina by Marcelo Magalhães
Edited by Karen Boss
Designed by Cathleen Schaad and Diane M. Earley
Production supervised by Jennifer Most Delaney

To all my math teachers, who showed me that numbers can be beautiful. —S. A.

To the time-traveling thinkers: Artemis, Mona, Niko, and Mallory—C. H.

Many thanks to the mathematicians and educators who reviewed this book, including Ted Heavenrich, Jennifer Irazabal, Melissa Guerrette, Rachel Harder, Stacy Mozer, Tim Wheeler, and Mostafa Mirabi. And an extra-special thanks to Dr. Keith Devlin, emeritus mathematician, Stanford University. And finally, thanks to sixth grader Korben W. for his helpful input.

Any errors are mine and not those of any expert reader. —S. A.